The LIBRARY of LANDFORMS™

# CANYONS

Isaac Nadeau

The Rosen Publishing Group's

**PowerKids Press**™

New York

*To Maged Nosshi*

Published in 2006 by The Rosen Publishing Group, Inc.
29 East 21st Street, New York, NY 10010

First Edition

Editor: Rachel O'Connor
Book Design: Elana Davidian

Photo Credits: Cover © Richard Cummins/Lonely Planet Images; p. 4 © Mark Daffey/Lonely Planet Images; p. 4 (inset) © Wade Eakle/Lonely Planet Images; p. 7 © Ralph Lee Hopkins/Lonely Planet Images; p. 8 (top) © Janice Marie Sheldon/Lonely Planet Images; p. 8 (bottom) © Layne Kennedy/Corbis; pp. 11 (left), 15 (right) © Jim Wark/Lonely Planet Images; p. 11 (right) © Richard Hamilton Smith/Corbis; p. 12 (top) © Donald C. & Priscilla Alexander Eastman/Lonely Planet Images; p. 12 (bottom left) © Gavin Hellier/naturepl.com; p. 12 (bottom right) © Marc Muench/Corbis; p. 15 (left) © Mark Newman/Lonely Planet Images; p. 16 © Gavin Anderson/Lonely Planet Images; p. 19 © Jeff Foott/naturepl.com; p. 20 (top left) © Kevin Fleming/Corbis; p. 20 (top right) © Tom Vezo/naturepl.com; p. 20 (bottom) © Mickey Gibson/Animals Animals.

Library of Congress Cataloging-in-Publication Data

Nadeau, Isaac.
  Canyons / Isaac Nadeau.— 1st ed.
     p. cm. — (Library of landforms)
  Includes index.
  ISBN 1-4042-3123-4
  1. Canyons—Juvenile literature. I. Title. II. Series.
  GB562.N33 2006
  551.44'2—dc22

                              2004025407

Manufactured in the United States of America

# CONTENTS

| | | |
|---|---|---|
| 1 | What Is a Canyon? | 5 |
| 2 | The Rocky Planet | 6 |
| 3 | Deposition | 9 |
| 4 | How Erosion Shapes Canyons | 10 |
| 5 | The Grand Canyon | 13 |
| 6 | Slot Canyons | 14 |
| 7 | Ocean Canyons | 17 |
| 8 | The Three Stages of a River Valley | 18 |
| 9 | Life in the Grand Canyon | 21 |
| 10 | People and Canyons | 22 |
| | Glossary | 23 |
| | Index | 24 |
| | Web Sites | 24 |

The Colca Canyon in Peru is one of the world's deepest canyons. It measures 10,509 feet (3,203 m) deep. It is also one of the longest, at a length of 63 miles (101 km). *Inset:* Here you can see the Yellowstone River as it rushes through the Grand Canyon in Yellowstone National Park in Wyoming.

# What Is a Canyon?

A canyon is a type of valley with steep cliffs for walls. Most canyons are formed by rivers or streams cutting downward through layers, or levels, of rock. Some canyons, called slot canyons, are very narrow. Their walls may be only a few feet (m) apart, but the canyon can be more than 100 feet (30 m) deep. Most canyons, however, like the Grand Canyon in Arizona, are much wider at the top than they are near the bottom. This is in part because **weathering** and **erosion** have had more time to wear away the tops of the walls than the recently uncovered bottoms of the walls.

Canyons can be found all over the world, and no two canyons are alike. Each canyon is formed under different conditions. The type of rock, the amount of rainfall, and the speed of the stream or river all play a part in creating a canyon. As a result each canyon has its own shape and size.

*The United States has many wonderful canyons in its landscape. One of the most scenic, or beautiful, is the Grand Canyon in Yellowstone Park. The canyon has been carved, or cut out, by the Yellowstone River, which flows for 24 miles (39 km) between the canyon cliffs.*

# THE ROCKY PLANET

A canyon is usually formed by a fast-moving river or stream flowing over rock. Without rock there could be no canyons. Earth is a rocky planet. From its hot core, or center, to the top of its highest mountains, Earth is made of rock. Since pieces of rock first began to cool on Earth's **molten** surface about 4.4 billion years ago, it has never stopped changing. Old rock is broken down by weathering and erosion, and new rock is formed by **deposition** of tiny bits of rock called **sediment** and by the cooling of magma, or hot liquid, from **volcanoes**.

There are three basic types of rock. Igneous rock, like basalt, is formed when molten rock cools and hardens. Sedimentary rock is formed when tiny bits of rock are deposited in layers, one on top of the other, and pressed together. Sandstone is a type of sedimentary rock. In some cases igneous or sedimentary rock is put under such high heat and pressure, or force, that the rock changes. This third type of rock is called metamorphic rock. **Marble** is a type of metamorphic rock that is formed from **limestone**. Some canyons are cut from all three types of rock.

Fossils are the remains, or traces, of animals or plants that died a long time ago. Fossils are often found in rocks. The fossil shown here is embedded in sedimentary rock. This shell probably dates from the Mesozoic age, which began about 245 million years ago.

| 3.5 BILLION YEARS AGO | 360-286 MILLION YEARS AGO | 65 MILLION YEARS AGO | 4-6 MILLION YEARS AGO | 1.8 MILLION YEARS AGO | 11,000 YEARS AGO | PRESENT |
|---|---|---|---|---|---|---|
| First proof of life is found in tiny fossils of bacteria in ancient rock. Bacteria are tiny living things that you can not see. | Much of the coal found in the eastern United States was formed. | A large-scale extinction, or dying out, occurs on Earth. Many believe this was caused by a meteor, or rock from outer space, hitting Earth. | The Grand Canyon begins to be carved by the Colorado River. | Pleistocene Epoch begins, marked by long periods of cold. About one-third of Earth is covered with ice sheets. | The most recent cold period ends, and the ice sheets withdraw to the poles. | All the forces that shape the land, including wind, water, waves, and living things, continue to change the face of Earth. |

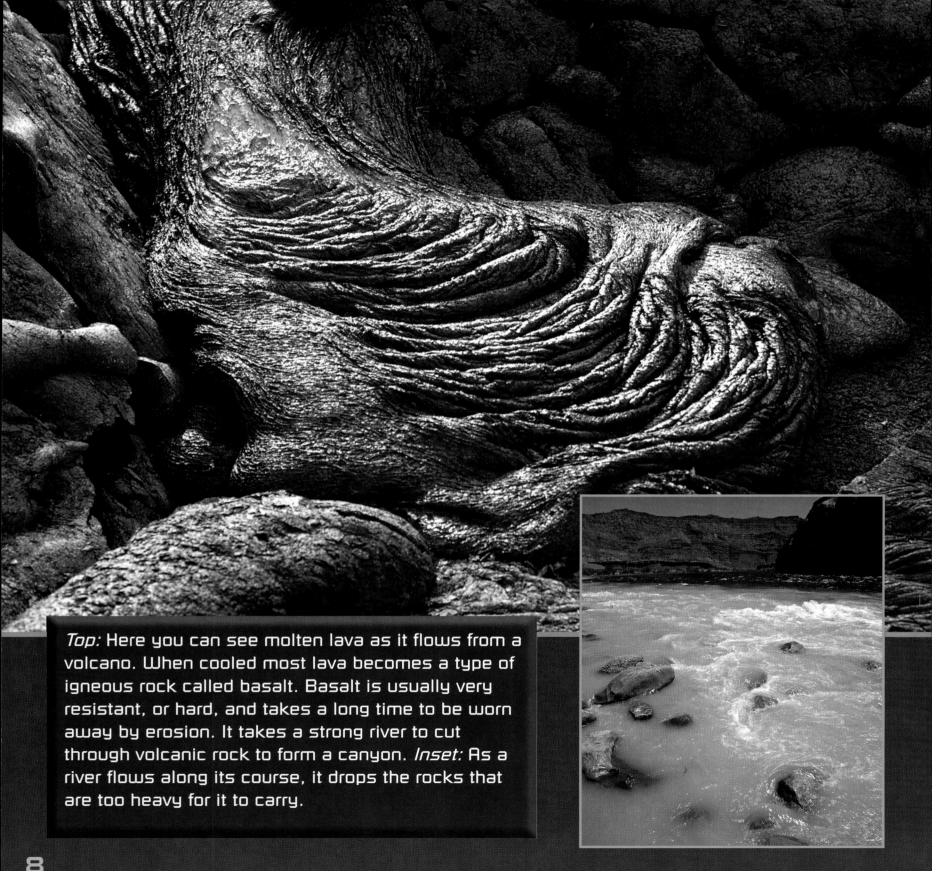

*Top:* Here you can see molten lava as it flows from a volcano. When cooled most lava becomes a type of igneous rock called basalt. Basalt is usually very resistant, or hard, and takes a long time to be worn away by erosion. It takes a strong river to cut through volcanic rock to form a canyon. *Inset:* As a river flows along its course, it drops the rocks that are too heavy for it to carry.

# Deposition

Deposition and erosion are important processes, or actions, in the formation of all landforms, including canyons. Every landform is the result of sediment being piled up or worn away. A canyon is the result of both deposition and erosion. Rocks must be deposited before erosion of the canyon can take place. Rock is deposited in many different ways. Some rock is deposited at the bottom of the ocean. When the sea level drops, this rock is exposed, or uncovered, to air and the forces of erosion. Many canyons are formed in rock that was once under water. Some rock is brought to Earth's surface as **lava** bubbling from the tops of volcanoes. Other rock is formed from the **dunes** of ancient deserts, where the sand, mud, and other sediments are pressed together to form sedimentary rock. In some places sediment is deposited more quickly than it is eroded, and layers of rock are piled on top of one another. The oldest rocks are at the bottom layer, and the youngest rocks are at the top. In many canyons you can see layers of different types of rock. Some of the bottom layers might have been deposited billions of years ago.

# How Erosion Shapes Canyons

*Resistant rocks, such as sandstone, wear away more slowly than softer rocks, such as slate. Softer rocks tend to form gentle slopes, while harder rocks tend to form cliffs. As the water cuts the rock more deeply, these cliffs become taller, forming the walls of a canyon.*

Water and **gravity** are two of the forces at work in the erosion of a canyon. Because of the pull of gravity, the water in rivers or streams flows downhill. The sediment in the water flows downhill with it. The sediment acts like sandpaper, scratching, or cutting, the ground and breaking off more sediment. Over time the river cuts deeper into the rock, forming a **channel**. Water and sediment continue to wear away at the floor and sides of the canyon, making it grow deeper, wider, and longer.

Weathering is another process that helps form canyons. **Chemical** weathering happens when the chemicals found in rain and river water dissolve rock, or wash it away. **Physical** weathering happens when water freezes in the cracks in rock, causing the rock to break. Physical weathering also occurs when plant roots grow inside rock and break it apart. Both chemical and physical weathering help form and shape canyons.

Marble Canyon is located in the Grand Canyon National Park. Here you can see another example of the Colorado River carving its way through the landscape. *Inset:* Physical weathering happens when rainwater flows into cracks or spaces in the rock. Sometimes the cold weather causes the water to freeze and when water freezes, it expands, or grows bigger. This expansion causes the rock to break apart.

*Top:* Here the Colorado River flows through the North Rim of the Grand Canyon. *Bottom Left:* Muleshoe Bend at the Colorado River runs through Glen Canyon, one of the river's smaller canyons. *Bottom Right:* Here you see the limestone walls of a narrow part of the Grand Canyon.

# THE GRAND CANYON

   Each year thousands of people from all over the world come to Arizona to visit the Grand Canyon, the most famous of all canyons. The Grand Canyon includes one main canyon, cut by the Colorado River, and hundreds of side canyons cut by its **tributaries**. The Colorado River begins high in the Rocky Mountains. The water rushes down the mountains, gathering sediment and picking up speed as it flows onto the Colorado Plateau. The Grand Canyon is cut into the rocks of the Colorado Plateau. A plateau is a large, flat area of high land. The Colorado Plateau covers about 13,000 square miles (33,670 sq km) in Arizona, New Mexico, Colorado, and Utah. There are many rivers on the Colorado Plateau, and they have cut many valleys and small canyons. The Grand Canyon is the biggest. The Grand Canyon is 277 miles (446 km) long. At its deepest it is about 6,000 feet (1,829 m) deep. At its widest it is more than 15 miles (24 km) wide. In the last four to six million years, the Colorado River has cut through rock that was laid down over the course of almost one billion years.

# Slot Canyons

Imagine standing at the bottom of a canyon where the walls are so close together that you can reach out and touch both of them at the same time. Sunlight can barely shine through the canyon's opening. This is what it is like to explore, or search inside, a slot canyon. Slot canyons begin as cracks in rock such as sandstone. They are formed by the work of wind and water. They are usually found in areas that get short bursts of rain a few times per year. Many slot canyons are dry most of the year. When it rains all the water uphill from the canyon flows over the land and into the canyon. The narrow areas of the slot canyon fill with this water quickly. When a canyon fills quickly with rainwater in this way, it is called a flash flood. It takes many years of flash flooding to cut a slot canyon, and they are quite rare, or uncommon.

As water flows through a slot canyon, it runs against both hard and soft stone. The softer stone wears away more quickly, creating beautiful patterns in the canyon walls. Wind, like water, carries tiny pieces of sand, brushing them against the walls of the canyon and smoothing them out.

Slot canyons can be found in the southwestern United States. One such canyon is the Upper Antelope Canyon in Arizona, pictured here. Much of the sandstone in the southwest was formed from sand dunes in an ancient desert millions of years ago. *Inset:* Here you can see an overhead view of a slot canyon off Lake Powell in Utah.

The Canyon-Dahab in the Middle East is very popular among divers. The mouth of this underwater canyon is about 49 feet (15 m) wide. It then narrows before you reach the widest part of the canyon, which measures about 82 feet (25 m).

# Ocean Canyons

Not all of Earth's canyons are found on land. There are also submarine, or underwater, canyons on the ocean floor. Submarine canyons are found just off the coasts of **continents**, where the land slopes beneath the water. Where the land dips beneath the ocean, the shore is continually pounded by waves. The waves help break down the rock, creating sediment. This sediment slides down the slope, like sand down the sides of a sand castle. As the sediment slides downhill, it scratches and wears down the surface of the underwater rock and carves out a canyon. Like canyons on land, submarine canyons begin as cracks or weak spots in the rock. The movement of water and sediment helps widen the cracks. The sediment then sinks to the floor near the bottom of the canyon. Submarine canyons can be many miles (km) long. Over time the sediment from shore is washed down the underwater channel and far out to sea. One of the biggest submarine canyons discovered so far is the Great Bahama Canyon in the Atlantic Ocean near the Bahamas. Its walls rise 14,060 feet (4,285 m) from the bottom to the top.

# The Three Stages of a River Valley

Canyons are part of river valleys. A river valley can be separated into three stages. The first stage occurs near the top of a mountain, where the hills are steep and the river is moving quickly. This is where canyons are formed because the water is moving quickly and cutting downward into the rock. At this stage the valley is said to be young. The canyon is said to be young because the land is steep and most of the erosion occurs at the bottom of the channel. At its next stage, the river valley is older, or mature. The land is less steep, and the river runs more slowly than before. In a canyon the amount of downward cutting lessens as the land becomes flatter. There is more erosion to the sides of the canyon. Over time the canyon is cut so wide that it stops being a canyon. A wide valley is left with water moving slowly through it. The last stage of a river valley is where the land is almost completely flat. At this stage the valley is said to be in its old age.

Waterfalls are usually found in the younger parts of a canyon, where the cliffs are steep and the water is running quickly. This waterfall in Stone Creek is a 3-mile (4.8 km) hike from the Colorado River.

*Top Left:* There are about 42 kinds of cacti found in Arizona's Grand Canyon. A cactus is able to last through long dry periods by storing water in its stem. *Top Right:* Here a white winged dove sits on a branch in the Grand Canyon. *Bottom:* A mule deer stands in the greenery of the North Rim of the Grand Canyon.

# LIFE IN THE GRAND CANYON

The Grand Canyon has many different habitats. A habitat is a place where a plant or an animal can find all the things it needs to live. Every living thing on Earth has special needs. Sunlight, water, **shelter**, and the right kinds of food are examples of these needs. Canyons can provide a home for many different kinds of creatures. Near the rim, or top, of the Grand Canyon, for example, **temperatures** are usually much colder than they are at the bottom. There is a lot of snow here in winter, and evergreen trees, such as spruce and fir, are common. Elk and mountain lions are among the animals that can be found near the canyon's rim. As you go deeper into the canyon and the temperatures get warmer, the kinds of plants and animals you see are different. Near the middle of the canyon there are junipers, scrub oaks, and pinyon pines, providing food and shelter for the pinyon mouse, cliff chipmunk, and mule deer. By the time you reach the bottom of the canyon, you are in a place much more like a desert, with yucca, cactus, and rattlesnakes. Temperatures reach above 100 degrees F (38° C) in summer at the bottom of the canyon.

# PEOPLE AND CANYONS

*There are several dams along the course of the Colorado River, providing electricity to many people living in the Southwest. In addition, much of the water from the Colorado River is channeled away to provide drinking water for cities.*

People visit canyons for many reasons. Some people like to enjoy the natural beauty of the canyon. Others go to canyons for adventure, looking for new rivers to raft and new walls to climb. **Geologists** explore canyons for clues about Earth's history.

Canyons, and the rivers that flow in them, also provide people with a **source** of electricity. People have controlled some of the rivers' power by building dams and trapping the rivers' energy, or force. In recent years, however, people have become concerned about the harm that dams and other human activities might do to canyons. In some cases dams are being removed to let the rivers run freely. More and more visitors to canyons and wild rivers are careful to leave them as they found them. In this way we can help make sure that all people who come to visit canyons will see the same wonderful creatures and formations we see in them today.

# Glossary

**channel** (CHA-nul)  The bed of a stream or a river.

**chemical** (KEH-mih-kul)  Having to do with matter that can be mixed with other matter to cause changes.

**continents** (KON-tih-nents)  The seven great masses of land on Earth.

**deposition** (deh-puh-ZIH-shun)  The dropping of tiny bits of rock in a new place.

**dunes** (DOONZ)  Hills of sand piled up by the wind.

**erosion** (ih-ROH-zhun)  The wearing away of land over time.

**geologists** (jee-AH-luh-jists)  Scientists who study the form of Earth.

**gravity** (GRA-vih-tee)  The force that causes objects to move toward the center of Earth.

**lava** (LAH-vuh)  A hot liquid made of melted rock that comes out of a volcano during an eruption.

**limestone** (LYM-stohn)  A kind of rock made of the bodies of small ocean animals.

**marble** (MAR-bul)  A hard, smooth stone used for buildings.

**molten** (MOL-ten)  Made liquid by heat.

**physical** (FIH-zih-kul)  Having to do with what something is made of.

**sediment** (SEH-dih-ment)  Gravel, sand, silt, or mud carried by wind or water.

**shelter** (SHEL-ter)  A place that guards someone from weather or danger.

**source** (SORS)  Where something comes from.

**temperatures** (TEM-pruh-cherz)  Measures of how hot or cold something is.

**tributaries** (TRIH-byoo-ter-eez)  Rivers or streams that flow into a larger river.

**volcanoes** (vol-KAY-nohz)  Openings on Earth's surface that sometimes shoot up lava.

**weathering** (WEH-thur-ing)  The breaking up of rock by water, wind, and chemical forces.

# INDEX

**C**
channel, 10
Colorado Plateau, 13
Colorado River, 13

**D**
deposition, 9
dunes, 9

**E**
erosion, 5–6, 9–10, 18

**G**
Grand Canyon, 5, 13, 21
gravity, 10
Great Bahama Canyon, 17

**I**
igneous rock, 6

**L**
lava, 9
limestone, 6

**M**
magma, 6
marble, 6
metamorphic rock, 6

**R**
river(s), 5–6, 10, 18, 22
Rocky Mountains, 13

**S**
sandstone, 6, 14
sediment, 6, 9–10, 17
sedimentary rock, 6
slot canyons, 5, 14
stream(s), 5–6, 10
submarine canyons, 17

**V**
valley(s), 18

**W**
weathering, 5–6, 10

# WEB SITES

Due to the changing nature of Internet links, PowerKids Press has developed an online list of Web sites related to the subject of this book. This site is updated regularly. Please use this link to access the list:
www.powerkidslinks.com/liblan/canyons/

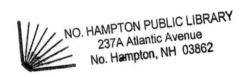